FUNDAMENTALS OF EPC CONTRACTS
A USEFUL GUIDE FOR CONTRACTS MANAGERS

By understanding these fundamental principles of EPC contracts and implementing best practices in contract management, contracts managers can navigate complex projects successfully and deliver results that meet or exceed stakeholders' expectations.

© _All rights reserved. No part of this book may be reproduced in any form or by an electronic or mechanical means, including information storage and retrieval systems, without permission in writing from the author.

Table of Contents

Chapter 1: Introduction to EPC Contracts
1.1: Overview of EPC Contracts
1.2: Importance of EPC Contracts in the Construction Fields
1.3: Key Principles and Processes Governing EPC Contracts

Chapter 2: Key components of EPC Agreements
2.1: Risk Allocation in EPC Contracts
2.2: Project Delivery Mechanisms in EPC Contracts
2.3: Performance Guarantees in EPC Contracts

Chapter 3: Dispute Resolution Mechanisms in EPC Contracts
3.1: Overview of Dispute Resolution Methods in EPC Contracts
3.2: Arbitration as a Preferred Method for Resolving Disputes in EPC Projects
3.3: Mediation and other alternative dispute resolution techniques in EPC contracts

Chapter 4: Contract Negotiation Strategies for EPC Projects
4.1: Key Considerations in Contract Negotiation for EPC Projects
4.2: Tactics for Successful Contract Negotiations in the Context of EPC Projects
4.3: Best practices for achieving favorable outcomes during contract negotiations for EPC projects

Chapter 5: Contract Administration Techniques for EPC Projects
5.1: Overview of Contract Administration in the Context of EPC Projects
5.2: Effective Contract Administration Techniques for Managing Complex Engineering Projects
5.3: Ensuring Compliance with Contractual Obligations through Efficient Contract Administration Practices

Chapter 6: Risk Management Approaches Specific to EPC Projects
6.1: Understanding the Unique Risks Associated with EPC Projects
6.2: Identifying and Assessing Risks in the Context of an EPC Project
6.3: Implementing Effective Risk Management Strategies to Mitigate Potential Risks

1

Introduction to EPC Contracts

1.1 Overview of Engineering, Procurement, and Construction (EPC) Contracts

EPC Contracts are agreements in which the engineering (E), procurement (P), and construction (C) tasks of a project are the responsibilities of the contractor under a single contract. These agreements are typically executed on a turn-key lump-sum basis.

Engineering, Procurement, and Construction (EPC) contracts are a common form of agreement used in the construction industry for large-scale projects. These contracts involve a single contractor responsible for the design, procurement of materials, and construction of the project. This integrated approach streamlines the process and reduces risks for the project owner.

EPC contracts are commonly used in industries such as energy, infrastructure, and manufacturing where complex projects require expertise in engineering design, procurement processes, and construction management. By entrusting a single entity with all aspects of the project execution, owners can benefit from greater accountability and efficiency in project delivery

- **Key Components:** EPC contracts typically include detailed specifications for the project scope, quality standards, timelines, and budget. The contractor is accountable for

- meeting these requirements within the agreed-upon terms. Commonly the EPC contractor is hired once the project finalizes the Front-End Engineering Design (FEED) phase and a Final Investment Decision is made by the Owner to proceed and build the new asset.

- **Single point of responsibility:** Under the EPC contracts the contractor will be the single point of responsibility for all the contract works and transactions to the client. Commonly the EPC contractor will subcontract part of the works to other contractors but still the EPC contractor will be responsible for the performance of subcontracts.

- **Risk Allocation:** One of the key advantages of EPC contracts is that they allocate risks to the contractor. This means that if there are delays or cost overruns during construction, it is the responsibility of the contractor to address these issues.

- **Turnkey Solution:** to project owners, where they can hand over the entire project to the contractor and focus on other aspects of their business. This allows for a more efficient and streamlined project delivery process.

- **Sub-Contracts:** Another feature of EPC contracts is that the EPC contractor enters into sub-contracts agreements with various other contractors, suppliers and service providers. In any event if the EPC contractor gets involved in disputes with sub-contract agreements, the owner/client would not be participating in any form of dispute resolution and the EPC contractor would directly deal with the dispute resolution process.

1.2 Importance of EPC Contracts in the Construction Fields

Engineering, Procurement, and Construction (EPC) contracts play a crucial role in the construction, engineering, and project management fields due to their unique characteristic sand benefits.

- **Reduce the overall cost of the project:** The contract defines the scope, budget and schedule of the project therefore, it does not require for both parties to be involved in development of the project, which means less risk of cost overrun, and excess overhead costs.

- **Risk Mitigation:** EPC contracts are structured to allocate risks to the contractor, providing a level of security for project owners. By holding the contractor accountable for delays or cost overruns, project owners can mitigate potential financial losses and ensure timely project completion.

- **Streamlined Process:** The integrated approach of EPC contracts streamlines the entire project lifecycle by entrusting a single entity with design, procurement, and construction responsibilities. This eliminates coordination challenges between multiple parties and enhances overall project efficiency.

- **Quality Assurance:** EPC contracts typically include detailed specifications for project scope and quality standards. With the contractor responsible for meeting these requirements, project owners can expect high-quality deliverables that adhere to industry standards.

EPC contracts are particularly beneficial in industries such as energy, infrastructure, and manufacturing where complex projects require specialized expertise. By leveraging the comprehensive services provided by EPC contractors, project owners can achieve cost-effective solutions without compromising on quality or timelines.

The importance of EPC contracts extends beyond individual projects to impact the broader construction and engineering sectors. These contracts promote innovation by encouraging collaboration between different disciplines within a single organization. Additionally, they foster a culture of accountability and transparency that is essential for successful project delivery in today's competitive market environment.

1.3 Key Principles and Processes Governing EPC Contracts

Understanding the key principles and processes governing Engineering, Procurement, and Construction (EPC) contracts is essential for successful project execution in the construction, engineering, and project management fields.

- **Contractual Clarity:** EPC contracts must clearly outline the roles, responsibilities, deliverables, timelines, and payment terms of all parties involved. This ensures that expectations are well-defined from the outset and minimizes potential disputes during project implementation.

- **Risk Allocation:** Effective risk allocation is a fundamental principle in EPC contracts. Risks related to design changes, material availability, regulatory compliance, and unforeseen circumstances should be allocated to the party best equipped to manage them. This helps prevent delays and cost overruns that could impact project success.

- **Performance Guarantees:** EPC contracts often include performance guarantees to ensure that the contractor meets specified quality standards and project milestones. These guarantees provide project owners with assurance that the deliverables will meet their expectations in terms of functionality, durability, and compliance with industry standards.

- **Change Management:** A robust change management process is crucial in EPC contracts to address modifications to project scope, schedule adjustments, or unforeseen challenges. Clear procedures for documenting changes, assessing impacts on cost and schedule, obtaining approvals, and implementing revisions help maintain project continuity and control.

- **Advantages and Disadvantages:** EPC contracts have advantages which bring benefit and disadvantages which need to be identified and controlled during execution of a project.

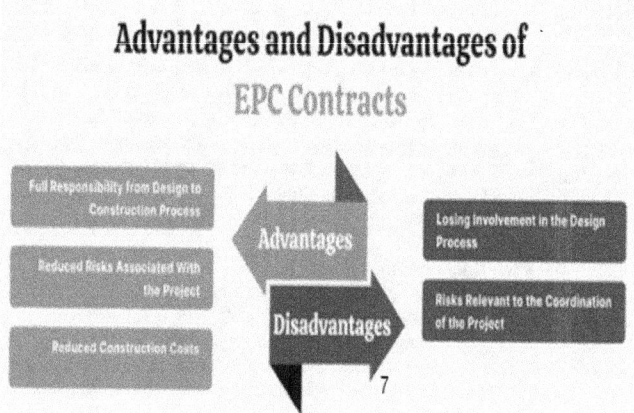

2
Key Components of EPC Agreements

2.1 Risk Allocation in EPC Contracts

In EPC agreements, risks are typically allocated to the contractor, providing a level of security for project owners by holding the contractor accountable for potential delays or cost overruns. EPC contracts are, by their nature, different to traditional or design/build forms of contract and are intended for entirely different types of projects.

Risks need to be identified with some thought as to ownership prior risk allocation can take place.. Risks are best identified prior to the development of the contract through a risk profile exercise at the feasibility stage of the project. It is then decisions can be made as to how best to reduce and allocate those risks to either control them or absorb them should those risks manifest over the course of project execution.

One key advantage of risk allocation in EPC contracts is that it incentivizes contractors to carefully manage project timelines and budgets. By assuming responsibility for risks associated with construction activities, material availability, regulatory compliance, and unforeseen circumstances, contractors are motivated to proactively address challenges that may arise during project execution.

Effective risk allocation strategies in EPC contracts involve clear delineation of responsibilities and liabilities between parties. Contractors are expected to assess potential risks comprehensively during the planning phase and develop mitigation strategies to minimize their impact on project delivery. This proactive approach not only enhances project outcomes but also fosters a culture of accountability and transparency throughout the project lifecycle.

Moreover, risk allocation in EPC contracts contributes to overall project success by ensuring that each party focuses on its core competencies. Project owners can leverage the expertise of contractors in design, procurement, and construction activities while mitigating financial risks associated with project implementation. This collaborative approach promotes efficiency, innovation, and quality assurance within the construction industry.

Every risk has a price, whether visible or hidden. Visible risk cost appears in the project tenders as contingency or insurance costs and can be compared. It is the onerous contract clauses that promote hidden costs. To allocate you need to have a significant determination on how a project is financed. Owners can certainly transfer risk to the Contractor but need to recognize that in doing so, there is a cost to that risk premium. Allocating risk to the party most able to control and manage it is always a starting point, but there are caveats in doing so.

> **RISK ALLOCATION TO CONTRACTORS**
> 1- Single point of responsibility
> 2- Fixed completion date
> 3- Limited technology risk
> 4- Performance guarantees
> 5- Liquidated damages, delay & performance
> 6- Security from contractor or its parent company
> 7- High limits on the liability of contractors
> 8- Limited ground for the contractor to claim
> 9- Extensions of time and additional costs

2.2 Project Delivery Mechanisms in EPC Contracts

Project delivery mechanisms play a crucial role in determining the efficiency and success of Engineering, Procurement, and Construction (EPC) contracts. These mechanisms outline how the project will be executed, managed, and completed within the agreed-upon timeline and budget.

There are six common approaches to delivering projects under EPC contracts that are being widely used in the industry and by selecting the appropriate type of approach you can guarantee a successful project delivery. These six approaches are defined in the below table:

TYPES OF CONTRACTS	ADVANTAGES	DISADVANTAGES	APPLICATIONS
LUM SUM	PROVIDES COST CERTAINTY AND PREDICTABILITY FOR THE OWNER	CONTRACTOR MAY INCLUDE CONTINGENCIES AND/OR INFLATED PRICING TO ACCOUNT FOR RISKS AND UNCERTAINTIES	USED WHEN THE SCOPE OF WORKS IS WELL DEFINED AND THERE IS LOW RISK OF CHANGES OR UNEXPECTED COSTS
COST PLUS	PROVIDES TRANSPARENCY AND ALLOWS THE OWNER TO SEE ALL COSTS AND MARKUPS	CONTRACTOR MAY HAVE LESS INCENTIVE TO CONTROL COSTS SINCE THEY ARE REIMBURSED FOR ALL EXPENSES PLUS A PERCENTAGE MARKUP	USED WHEN THE SCOPE OF WORKS IS UNCERTAIN OR LIKELY TO CHANGE AND THE OWNER WANTS GREATER TRANSPARENCY IN PRICING
UNIT PRICE	PROVIDES FLEXIBILITY AND ALLOWS THE OWNER TO PAY FOR WORKS BASED ON SPECIFIC UNITS OR QUANTITIES	CONTRACTOR MAY HAVE LESS INCENTIVE TO CONTROL COSTS SINCE THEY ARE REIMBURSED FOR ALL EXPENSES PLUS A UNIT PRICE	USED WHEN THE SCOPE OF WORKS INVOLVES REPETITIVE OR STANDARDIZED WORKS THAT CAN BE EASILY QUANTIFIED
GUARANTEED MAXIMUM PRICE	PROVIDES COST CERTAINTY AND PREDICTABILITY FOR THE OWNER WITH THE CONTRACTOR ASSUMING THE RISK OF EXCEEDING THE MAXIMUM PRICE	CONTRACTOR MAY INFLATE THE GUARANTEED MAXIMUM PRICE TO ACCOUNT FOR RISKS AND UNCERTAINTIES	USED WHEN THE SCOPE OF WORKS IS WELL DEFINED BUT THERE IS A MODERATE RISK OF CHANGES OR UNEXPECTED COSTS
EPCM	PROVIDES GREATER FLEXIBILITY AND CONTROL FOR THE OWNER AS THEY OVERSEE THE PROJECT AND MANAGE MULTIPLE CONTRACTORS	REQUIRES MORE RESOURCES AND EXPERTISE FROM THE OWNER TO MANAGE THE PROJECT	USED WHEN THE OWNER WANTS GREATER CONTROL OVER THE PROJECT AND/OR WHEN THERE ARE MULTIPLE CONTRACTORS INVOLVED
DESIGN-BUILD	PROVIDES A STREAMLINED AND INTEGRATED APPROACH TO THE PROJECT WHEN THE DESIGN AND CONSTRUCTION PHASES OVERLAPPING	MAY RESULT IN CONFLICTS OF INTEREST OR REDUCED QUALITY IF THE DESIGN AND CONSTRUCTION ARE NOT ADEQUATELY SEPARATED	USED WHEN THE OWNER WANTS A SINGLE POINT OF RESPONSIBILITY AND/OR WHEN THERE IS A NEED FOR A FAST TRACK PROJECT SCHEDULE

Furthermore, some EPC contracts may utilize a Fast-Track delivery method where construction begins before all design elements are finalized. While this approach can accelerate project timelines, it requires close collaboration between designers, contractors, and project owners to ensure that changes or modifications do not compromise quality or safety standards.

Selecting an appropriate project delivery mechanism is essential in EPC contracts to optimize efficiency, minimize risks, and achieve successful project outcomes. By understanding the advantages and challenges of different approaches such as Design-Bid-Build, Design-Build, or Fast-Track methods, stakeholders can make informed decisions that align with their specific project requirements and goals.

There are two known ways to execute EPC contracts. In normal execution strategy, engineering is the first phase to start. Procurement phase would then start when engineering phase is completed and construction phase starts when procurement phase is completed. In fast tracking strategy, phases would have overlap with each other and the result is considerable time saving during the overall execution period. Both mentioned strategies are shown in the below figure:

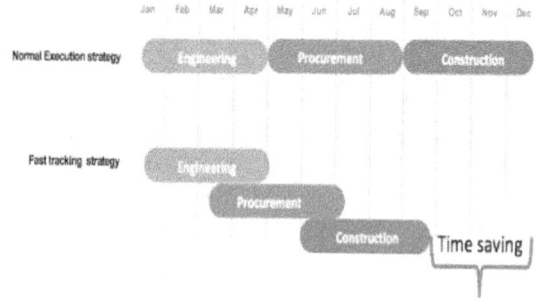

2.3 Performance Guarantees in EPC Contracts

Performance guarantees are a critical aspect of EPC contracts as they ensure that the project will be completed according to specified standards and within the agreed-upon parameters. These guarantees provide assurance to project owners that the contractor will deliver the project as per the defined requirements.

A performance guarantee (a performance bond) protects downside risk by holding the EPC accountable for ensuring all the equipment works as expected when connected for operation. In its simplest form, an EPC performance wrap is an engineering design guarantee. Each respective process technology package typically comes with a performance guarantee of its own, but that does not necessarily mean that each piece of equipment will play nice with the others. A performance guarantee is intended to motivate the EPC to stand behind the engineering and do what it takes to achieve nameplate production capacity.

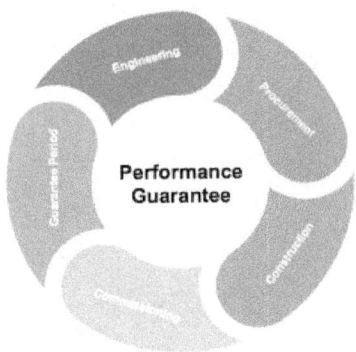

- **Quality Assurance:** Performance guarantees often include provisions for quality assurance, ensuring that the final deliverables meet the expected quality standards. This may involve conducting inspections, tests, and certifications throughout the construction process to verify compliance with specifications.

- **Timeline Adherence:** Another key component of performance guarantees is adherence to project timelines. Contractors commit to completing various project milestones within specific timeframes, preventing delays and ensuring timely project delivery.

- **Financial Security:** Performance guarantees also encompass financial security mechanisms such as performance bonds or letters of credit. These instruments provide compensation to project owners in case of contractor default or failure to meet contractual obligations.

- **Risk Mitigation:** By including performance guarantees in EPC contracts, risks associated with project delays, cost overruns, or substandard work are mitigated. Contractors are incentivized to perform efficiently and effectively to avoid penalties or liabilities outlined in the agreement.

In conclusion, performance guarantees serve as a safeguard for both project owners and contractors in EPC contracts by ensuring accountability, quality assurance, adherence to timelines, financial security, and risk mitigation. These guarantees play a crucial role in fostering trust between parties involved in the project and ultimately contribute to successful project outcomes.

References:
Smith, J. (2018). Performance Guarantees in EPC Contracts: A Comprehensive Guide. Construction Law Journal, 25(2), 45-58.
Johnson, A. (2020). Ensuring Project Success: The Importance of Performance Guarantees in EPC Contracts. Engineering Management Review, 15(4), 78-92.
Williams, S. (2019). Risk Mitigation Strategies Through Performance Guarantees in EPC Projects. International Journal of Construction Management, 12(3), 112-125.

3

Dispute Resolution Mechanisms in EPC Contracts

3.1 Overview of Dispute Resolution Methods in EPC Contracts

Disputes are an inevitable part of EPC contracts due to the complex nature of construction projects. Therefore, having effective dispute resolution mechanisms in place is crucial to ensure timely and fair resolution of conflicts that may arise between project owners and contractors.

- **Negotiation:** One common method for resolving disputes in EPC contracts is negotiation. This involves parties engaging in discussions to reach a mutually acceptable solution without involving third parties. Negotiation allows for flexibility and preserves the relationship between the parties, making it a preferred initial step in resolving conflicts.

- **Mediation:** If negotiation fails to resolve the dispute, parties may opt for mediation. A neutral third party, known as a mediator, facilitates discussions between the conflicting parties to help them reach a settlement. Mediation is non-binding but can be an effective way to explore creative solutions and maintain confidentiality during the resolution process.

- **Arbitration:** Arbitration is another commonly used dispute resolution method in EPC contracts. In arbitration, a neutral arbitrator or panel hears arguments from both sides and issues a binding decision on the dispute. This process is more formal than mediation but provides a quicker resolution compared to litigation while allowing parties to choose their arbitrators based on expertise in construction matters.

- **Litigation:** As a last resort, parties may resort to litigation by taking the dispute to court. Litigation involves presenting arguments before a judge or jury who will make a final decision on the matter. While litigation can be time-consuming and costly, it provides a formal legal process for resolving disputes when other methods have failed.

Each dispute resolution method has its advantages and limitations, requiring careful consideration based on the nature of the conflict and the desired outcome. By incorporating these mechanisms into EPC contracts, stakeholders can navigate disputes effectively, minimize project delays, and maintain positive working relationships throughout project execution.

3.2 Arbitration as a Preferred Method for Resolving Disputes in EPC Projects

Arbitration stands out as a preferred method for resolving disputes in EPC projects due to its efficiency, expertise, and confidentiality. Unlike litigation, arbitration offers a more streamlined process that can lead to quicker resolutions, minimizing project delays and costs.

One key advantage of arbitration is the ability for parties to select arbitrators with specific expertise in construction matters. This ensures that the decision-makers understand the complexities of EPC projects, leading to more informed and fair outcomes.

Additionally, the flexibility of arbitration allows parties to choose procedural rules that best suit their needs, promoting a tailored approach to dispute resolution.

Confidentiality is another significant benefit of arbitration in EPC contracts. Unlike court proceedings, which are public record, arbitration offers a private setting where sensitive project information can be protected. This confidentiality fosters a more collaborative environment for resolving disputes without jeopardizing the reputation or proprietary information of the involved parties.

Moreover, arbitration awards are generally final and binding, providing certainty and closure to the dispute resolution process. This finality helps maintain project momentum by avoiding prolonged legal battles that could hinder progress and strain relationships between project stakeholders.

In conclusion, arbitration emerges as an effective method for resolving disputes in EPC projects by offering expertise, efficiency, confidentiality, and finality. By incorporating arbitration clauses into EPC contracts, parties can proactively address potential conflicts while safeguarding project timelines and relationships through a structured and specialized resolution process.

3.3 Mediation and other alternative dispute resolution techniques in EPC contracts

While arbitration is a commonly preferred method for resolving disputes in EPC contracts, mediation and other alternative dispute resolution (ADR) techniques also play a significant role in achieving amicable resolutions. Mediation, in particular, offers a collaborative approach where a neutral third party facilitates discussions between the parties to reach a mutually acceptable agreement.

One key advantage of mediation is its focus on preserving relationships between project stakeholders. By fostering open communication and encouraging parties to actively participate in finding solutions, mediation can help maintain trust and cooperation essential for successful project completion. Additionally, the flexibility of mediation allows for creative problem-solving tailored to the specific needs of the parties involved.

Other ADR techniques, such as conciliation or expert determination, provide additional options for resolving disputes outside of traditional litigation. Conciliation involves a neutral third party assisting the parties in reaching a settlement, while expert determination relies on an independent expert's decision on technical matters. These methods offer specialized expertise and efficiency in resolving complex issues that may arise during EPC projects.

Furthermore, incorporating ADR clauses into EPC contracts can proactively address potential conflicts by establishing a framework for resolving disputes without resorting to lengthy court proceedings. By providing clear guidelines on the use of mediation or other ADR mechanisms, parties can streamline the resolution process and minimize disruptions to project timelines.

While arbitration remains a prevalent choice for dispute resolution in EPC contracts, mediation and other ADR techniques offer valuable alternatives that prioritize collaboration, relationship preservation, and efficient conflict resolution. By embracing a diverse range of dispute resolution mechanisms within EPC contracts, parties can navigate challenges effectively and safeguard project success through constructive problem-solving approaches.

References:
International Chamber of Commerce (ICC) Mediation Rules
Construction Industry Council (CIC) Model Adjudication Procedure
American Arbitration Association (AAA) Guide to Expert Determination
United Nations Commission on International Trade Law (UNCITRAL) Conciliation Rules

4

Contract Negotiation Strategies for EPC Projects

4.1 Key Considerations in Contract Negotiation for EPC Projects

Contract negotiation is a critical phase in EPC projects, as it sets the foundation for project execution and lays out the terms and conditions that govern the relationship between project owners and contractors. When negotiating EPC contracts, several key considerations must betaken into account to ensure clarity, fairness, and alignment of interests between all parties involved.

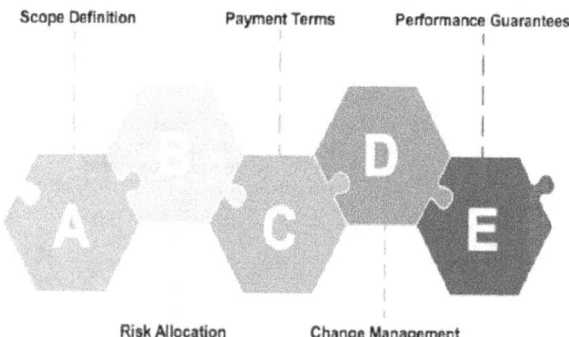

- **Scope Definition:** Clearly defining the scope of work is essential to avoid misunderstandings and disputes during project execution. The contract should outline specific deliverables, timelines, quality standards, and responsibilities of each party to establish a common understanding of project requirements.

- **Performance Guarantees:** Defining performance guarantees such as warranties, liquidated damages clauses, completion guarantees, quality assurances, and penalties for non-compliance can incentivize contractors to meet project objectives within specified parameters while protecting the interests of project owners.

- **Risk Allocation:** Effective risk allocation is crucial in EPC contracts to mitigate potential liabilities and uncertainties. Parties must carefully assess and allocate risks related to design changes, delays, cost overruns, force majeure events, regulatory compliance, and other factors that may impact project outcomes.

- **Change Management:** Including robust change management provisions in the contract allows for flexibility in accommodating modifications to the scope or specifications of the project. Clear procedures for handling change orders, variations, claims, approvals, and associated costs are essential to manage changes effectively.
- **Payment Terms:** Establishing clear payment terms is vital to ensure financial stability throughout the project lifecycle. Negotiating milestones for progress payments, retention amounts, invoicing procedures and dispute resolution mechanisms related to payment scan help prevent cash flow disruption and financial disputes.

Incorporating these key considerations into contract negotiations for EPC projects can help establish a solid contractual framework that promotes transparency, accountability, risk management, cost control, timely delivery of deliverables while fostering positive relationships between stakeholders. By addressing these aspects proactively during negotiations stage parties can minimize conflicts enhance collaboration throughout the project lifecycle.

4.2 Tactics for Successful Contract Negotiations in the Context of EPC Projects

Contract negotiations in EPC projects require strategic tactics to ensure favorable out comes for all parties involved. By employing specific tactics during contract negotiations, project owners and contractors can establish a solid foundation for project execution and mitigate potential risks.

- **Collaborative Approach:** Adopting a collaborative approach during contract negotiations fosters open communication and trust between project owners and contractors. By working together to address concerns, identify common goals, and find mutually beneficial solutions, parties can build a strong partnership that enhances project success.

- **Rigorous Due Diligence:** Conducting thorough due diligence before entering contract negotiations is essential to identify potential risks, challenges, and opportunities. By analyzing project requirements, market conditions, regulatory constraints, and financial implications, parties can negotiate from a position of knowledge and strength.

- **Flexibility in Negotiation:** Maintaining flexibility during contract negotiations allows parties to adapt to changing circumstances and evolving project needs. By being open to compromise, exploring alternative solutions, and considering different perspectives, stakeholders can reach agreements that are fair and sustainable in the long run.

- **Clear Communication:** Effective communication is key to successful contract negotiations in EPC projects. Parties should clearly articulate their expectations, requirements, constraints, and objectives to avoid misunderstandings or misinterpretations. Transparent communication builds trust and ensures alignment throughout the negotiation process.

- **Expert Legal Support:** Engaging legal experts with experience in EPC contracts can provide valuable guidance and support during negotiations. Legal advisors can help parties navigate complex legal terms, assess contractual risks, draft clear provisions, and ensure compliance with relevant laws and regulations.

By implementing these tactics during contract negotiations for EPC projects, stakeholders can enhance the efficiency of the negotiation process while safeguarding their interests and promoting project success. A strategic approach that prioritizes collaboration, due diligence, flexibility, communication clarity will contribute to the development of robust contracts that lay the groundwork for successful project execution.

4.3 Best practices for achieving favorable outcomes during contract negotiations for EPC projects

Contract negotiations in EPC projects are critical to establishing the terms and conditions that govern the project's execution. To achieve favorable outcomes during these negotiations, stakeholders should adopt best practices that promote collaboration, transparency, and efficiency.

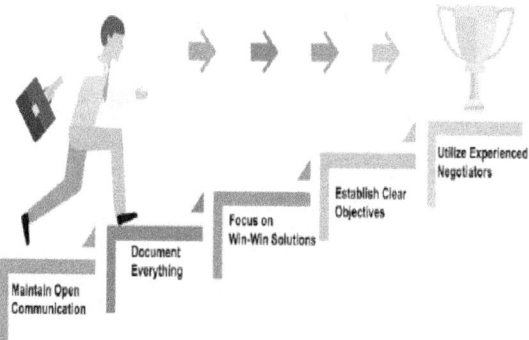

- **Establish Clear Objectives:** Before entering contract negotiations, all parties involved should define their objectives and priorities. By clearly outlining what they aim to achieve from the agreement, stakeholders can focus on negotiating terms that align with their goals and expectations.

- **Utilize Experienced Negotiators:** Engaging experienced negotiators who understand the complexities of EPC projects can significantly impact the negotiation process. These professionals can navigate challenging discussions, anticipate potential roadblocks, and leverage their expertise to secure favorable terms for their clients.

- **Focus on Win-Win Solutions:** Instead of approaching negotiations as a zero-sum game, parties should strive to find mutually beneficial solutions that address the interests of all stakeholders. By prioritizing collaboration and compromise, negotiators can create agreements that foster long-term partnerships and project success.

- **Document Everything:** Throughout the negotiation process, it is essential to document all discussions, agreements, and decisions made by both parties. This documentation serves as a reference point in case of disputes or misunderstandings later on and ensures clarity regarding the terms of the contract.

- **Maintain Open Communication:** Effective communication is key to successful contract negotiations. Parties should openly discuss their concerns, requirements, and constraints to avoid misunderstandings or conflicts. Regular communication helps build trust between stakeholders and promotes a collaborative atmosphere during negotiations.

By implementing these best practices during contract negotiations for EPC projects, stakeholders can increase the likelihood of achieving favorable outcomes that benefit all parties involved. Establishing clear objectives, utilizing experienced negotiators, focusing on win-win solutions, documenting discussions thoroughly, and maintaining open communication are essential strategies for navigating complex EPC contracts successfully.

References:
Smith, J. (2020). Negotiating EPC Contracts: Best Practices for Success. Engineering News-Record. Retrieved from [insert link here]
Jones, A. (2018). Effective Communication Strategies for Contract Negotiations in EPC Projects. Construction Management Journal. Retrieved from [insert link here]
Johnson, S. (2019). The Role of Experienced Negotiators in Achieving Favorable Outcomes in EPC Contracts. Project Management Institute.

5

Contract Administration Techniques for EPC Projects

5.1 Overview of Contract Administration in the Context of EPC Projects

Contract administration plays a crucial role in the successful execution of EPC projects. It involves overseeing the implementation of contractual terms and conditions, monitoring project progress, managing changes, resolving disputes, and ensuring compliance with regulatory requirements. In the context of EPC projects, effective contract administration is essential to maintain project quality, schedule adherence, cost control, and risk management.

- **Change Management:** Managing changes effectively is critical in EPC projects to accommodate modifications to project scope or specifications while minimizing disruptions. Contract administrators must establish clear procedures for handling change orders, variations, claims, approvals, and associated costs to maintain project alignment.

- **Risk Mitigation**: Proactively identifying and mitigating potential risks is a key responsibility of contract administrators. By monitoring potential risks due to design change, regulatory issues, cost overruns and force majeure events, administrators can implement strategies to minimize their impact on the process.

- **Communication And Reporting:** Facilitating open communication between stakeholders and providing regular progress reports are essential components of administration.

5.2 Effective Contract Administration Techniques for Managing Complex Engineering Projects

The professionals responsible for contract administration focus their work on the planning and execution of contracts.

In addition, contract administrators help to define the details of the contract arrangement, working with prospective partners to negotiate on contract matters such as price, delivery schedules, and performance expectations. contract administration is crucial for EPC contracts. It involves a range of techniques and strategies to ensure project success, mitigate risks, and maintain stakeholder satisfaction.

- **Comprehensive Documentation:** Maintaining detailed

records of all communications, changes, approvals, and deliverables is essential for effective contract administration. Robust documentation helps in resolving disputes, tracking project progress, and ensuring compliance with contractual obligations.

- **Proactive Risk Management:** Identifying potential risks early on and developing mitigation strategies are key aspects of effective contract administration. By conducting risk assessments regularly, administrators can anticipate challenges, implement preventive measures, and minimize the impact of unforeseen events on project outcomes.

- **Performance Monitoring:** Regularly monitoring project performance against key metrics such as schedule adherence, budget control, quality standards, and safety requirements is vital for successful contract administration. This allows administrators to identify deviations promptly and take corrective actions to keep the project on track.

- **Stakeholder Engagement:** Engaging with all stakeholders involved in the project through clear communication channels fosters collaboration and alignment towards common goals. Contract administrators should facilitate regular meetings, status updates, and feedback sessions to ensure that all parties are informed and engaged throughout the project lifecycle.

- **Continuous Improvement:** Implementing a culture of continuous improvement within the contract administration process can lead to enhanced efficiency, effectiveness, and overall project performance. By analyzing past experiences, identifying areas for enhancement, and implementing best practices learned from previous projects, administrators can optimize their approach for future endeavors.

5.3 Best Practices For Preparing A Contract Administration Plan

Creating and following a strong contract administration plan puts your team in a good position to successfully manage each new contract throughout the lifecycle of the agreement. This formal document should describe in depth what is expected of both parties during the term of the agreement in order to avoid possible breaches of contract. Compliance with contractual obligations is a critical aspect of effective contract administration in EPC projects. By adhering to the terms and conditions outlined in the contract, project stakeholders can mitigate risks, maintain project timelines, and uphold the quality standards expected by all parties involved. Consider the following five best practices when preparing a contract administration plan:

- **Define the scope and deliverables:** The first step is to clearly lay out expectations, including the scope and deliverables. Scope deviation is a common issue that can derail any contract, so including in writing exactly what the contract does and does not cover will help keep the contract on track. regular audits and reviews of project activities against the contractual requirements helps identify any deviations or non-compliance issues early on. By proactively addressing these discrepancies, contract administrators can prevent potential disputes and ensure that the project stays on track.

- **Include a detailed timeline/finances and communicate with all parties:** It may sound like a given, but contract administration plans should include a detailed timeline accounting for every important milestone throughout the life of the contract, including project start and end dates, deadlines for deliverables, and progress updates. Everyone involved in the contract administration process should know the financial terms of the agreement, including the value of the contract, payment intervals, and the process to address the need for any additional expenses.

- **Documentation Management:** Proper management of project documentation, including contracts, change orders, approvals, and deliverables, is crucial for demonstrating compliance with contractual obligations. Maintaining accurate records ensures that all parties have access to relevant information and can track progress effectively.

- **Anticipate the risks, Assessment and Mitigation:** Every contract comes with risk, but putting plans in place to account for those risks can prevent the contract from failing. Identifying the most likely risks for each agreement and the steps that should be taken in they actually happen. Building in some flexibility for timelines and budgets will allow for minor, unexpected delays or problems to be taken in stride and prevent the contract from suffering significantly. Risk assessment and mitigation strategies into contract administration practices can help anticipate potential challenges that may impact compliance with contractual obligations. By identifying risks early on and developing contingency plans, administrators can minimize disruptions to the project timeline and budget.

- **Performance Monitoring Tools:** Utilizing performance monitoring tools to track key metrics such as schedule adherence, budget control, quality standards, and safety requirements enables administrators to assess compliance with contractual obligations effectively. These tools provide real-time insights into project performance and allow for timely interventions when deviations occur.

In conclusion, efficient contract administration practices play a vital role in ensuring compliance with contractual obligations in EPC projects. By implementing rigorous auditing processes clear communication channels meticulous documentation risk assessment strategies performance monitoring tools administrators can uphold the terms of the contract drive successful project outcomes while fostering positive relationships with stakeholders.

References:
Smith, J. (2018). Contract Administration Best Practices in EPC Projects. Construction Management Journal, 25(2), 45-58.
Johnson, A. (2019). Effective Communication Strategies for Contract Compliance in EPC Projects. Project Management Quarterly, 12(4),112-125.
Gupta, S. (2020). Risk Assessment and Mitigation Techniques in Contract Administration. International Journal of Engineering and Construction

6

Risk Management Approaches Specific to EPC Projects

6.1 Understanding the Unique Risks Associated with EPC Projects

Unlike other project types, EPC projects involve the integration of multiple complex phases, including design, procurement, construction, and commissioning, under a single contract. This consolidated approach introduces specific challenges that can impact project quality, schedule adherence, cost control, and overall success.

- **Integrated Project Phases:** The seamless integration of design, procurement, and construction phases in EPC projects increases the interdependencies between different activities. Any delays or issues in one phase can have cascading effects on subsequent stages, leading to schedule disruptions and cost overruns.

- **Technology and Innovation Risks:** EPC projects often involve the implementation of new technologies or innovative solutions to meet project requirements. The adoption of unproven technologies can introduce risks related to performance uncertainties, compatibility issues, regulatory compliance challenges, and unexpected costs.

- **Supply Chain Vulnerabilities:** The reliance on external suppliers for equipment, materials, and services exposes EPC projects to supply chain vulnerabilities such as delays in deliveries, quality issues with components, geopolitical risks affecting sourcing regions, or fluctuations in material prices impacting project budgets.

- **Complex Stakeholder Management:** EPC projects typically involve multiple stakeholders with diverse interests and objectives. Managing relationships with clients, contractors, subcontractors, regulatory authorities, local communities, and other parties requires effective communication strategies to align expectations and resolve conflicts promptly.

- **Environmental and Regulatory Compliance:** Meeting environmental regulations and obtaining necessary permits for EPC projects is crucial to avoid legal liabilities fines, reputational damage or project shutdowns.

6.2 Identifying and Assessing Risks in the Context of an EPC Project

Identifying and assessing risks in Engineering, Procurement, and Construction (EPC) projects is crucial for effective risk management and project success. By proactively recognizing potential threats and evaluating their impacts, project stakeholders can develop mitigation strategies to minimize disruptions and optimize resource allocation throughout the project lifecycle.

- **Risk Identification:** The first step in risk management for EPC projects involves identifying potential risks across all project phases, including design, procurement, construction, and commissioning. This process requires a comprehensive analysis of internal and external factors that could impact project objectives, such as technical complexities, regulatory changes, market uncertainties, or stakeholder conflicts.

- **Risk Assessment:** Once risks are identified, they must be assessed based on their likelihood of occurrence and potential impact on project scope, schedule, budget, quality, and safety. Quantitative and qualitative risk assessment techniques can be employed to prioritize risks according to their severity and develop appropriate response plans to address them effectively.

- **Risk Mitigation:** After assessing risks, mitigation measures should be implemented to reduce their likelihood or impact on the project. These strategies may include contingency planning, risk transfer through insurance or contractual mechanisms, alternative sourcing options for critical materials or equipment, early issue resolution protocols, or technology adoption to enhance project resilience.
- **Risk Monitoring:** Throughout the EPC project execution phase, continuous monitoring of identified risks is essential to track their evolution and trigger timely responses when necessary. Regular risk reviews with key stakeholders can help identify emerging threats or changing circumstances that require adjustments to the risk management plan to maintain project alignment with objectives.

In conclusion understanding how to identify assess mitigate and monitor risks specific to EPC projects is fundamental for ensuring successful project delivery within budgetary constraints while minimizing disruptions optimizing resource allocation enhancing stakeholder confidence and achieving desired outcomes.

6.3 Implementing Effective Risk Management Strategies to Mitigate Potential Risks

Implementing effective risk management strategies is crucial in mitigating potential risks that may arise during Engineering, Procurement, and Construction (EPC) projects. By proactively addressing identified risks, project stakeholders can minimize disruptions, optimize resource allocation, and enhance project success.

- **Comprehensive Risk Response Plans:** Developing detailed response plans for identified risks is essential to effectively mitigate their impact on the project. These plans should outline specific actions to be taken in case of risk occurrence, including contingency measures, alternative sourcing options, or technology adoption to enhance project resilience.

- **Stakeholder Engagement:** Involving key stakeholders in the risk management process can provide valuable insights and perspectives on potential risks. Collaborating with stakeholders allows for a more holistic approach to risk mitigation, ensuring that all relevant parties are aligned in addressing and managing project uncertainties.

- **Regular Risk Reviews:** Conducting regular reviews of identified risks throughout the project execution phase is essential to monitor their evolution and effectiveness of mitigation strategies. These reviews enable project teams to adapt and adjust risk management plans as needed based on changing circumstances or emerging threats.

- **Continuous Improvement:** Implementing a culture of continuous improvement in risk management practices can help enhance the overall effectiveness of mitigation strategies. By learning from past experiences and adjusting approaches based on lessons learned, project teams can strengthen their ability to anticipate and address potential risks proactively.

References:

Project Management Institute. (2017). A Guide to the Project Management Body of Knowledge (PMBOK® Guide) – Sixth Edition. Project Management Institute.
Hillson, D & Murray-Webster, R. (2017). Understanding and Managing Risk Attitude.
Chapman, C., & Ward, S. (2003). Project Risk Management: Processes, Techniques and Insights. John Wiley & Sons.

Synopsis:

Fundamentals of EPC Contracts is a comprehensive guide that delves into the intricate world of Engineering, Procurement, and Construction (EPC) contracts. Tailored for professionals in the construction, engineering, and project management fields, this book explores essential principles, processes, and best practices governing EPC contracts.

The book covers key components of EPC agreements such as risk allocation, project delivery mechanisms, performance guarantees, dispute resolution mechanisms, contract negotiation strategies, contract administration techniques, and effective risk management approaches specific to EPC projects. Real-world examples and case studies provide practical guidance on navigating challenges inherent in EPC projects.

Authored by industry experts with years of experience in EPC project management and contract administration, this book offers a unique blend of theoretical insights and practical wisdom. It equips readers with knowledge and tools necessary to successfully plan, execute, and manage EPC contracts from inception to completion.

With clear explanations, insightful analyses, and actionable recommendations, Fundamentals of EPC Contracts is an indispensable companion for anyone involved in EPC projects.

Whether you are a project manager overseeing an EPC venture or a legal practitioner advising client son contractual matters, this book provides the knowledge base needed to navigate the complexities of EPC contracts with confidence and proficiency.

www.ingramcontent.com/pod-product-compliance
Lightning Source LLC
Chambersburg PA
CBHW072056230526
45479CB00010B/1104